Unwritten Laws of ...

ETHICS AND CHANGE IN ENGINEERING

TABLE OF CONTENTS

PART 1:

ETHICS FOR ENGINEERS

Doing the Right Thing When No One is Looking

THE HISTORY OF ENGINEERING ETHICS

INTRODUCTION

Part 1 of this book is designed to provide a clear picture of what ethical behavior looks like and ideas for what to do if you or someone else is not behaving in a manner consistent with this picture.

This is the first chapter in Part 1, entitled The History of Engineering Ethics.

This chapter is designed to give you a good feel for the history of engineering ethics – why and how they originated and how they have evolved over the years.

Read the following situation in which an engineer is faced with an ethical dilemma. After you read about the circumstances, consider these questions: What is the ethical dilemma? And what are the first 3 steps the engineer should take?

WHAT SHOULD PETER DO?

Peter has been working with the Ameral Oil Company's local affiliate for several years, and he has established a strong, trusting relationship with Jesse, manager of the local facility.

The facility, on Peter's recommendations, has followed all of the environmental regulations to the letter, and it has a solid reputation with the state regulatory agency. The local facility receives various petrochemical products via pipelines and tank trucks, and it blends them for resale to the private sector.

Jesse has been so pleased with Peter's work that he has recommended that Peter be retained as the corporate consulting engineer. This would be a significant advancement for Peter and his consulting firm, cementing Peter's steady and impressive rise in the firm. There is talk of a vice presidency in a few years.

One day, over coffee, Jesse starts telling Peter a story about a mysterious loss in one of the raw petrochemicals he receives by pipeline. Sometime during the 1970s, when operations were more lax, a loss of one of the process chemicals was discovered when the books were audited. There were apparently 10,000 gallons of the chemical missing. After running pressure tests on the pipelines, the plant manager found that one of the pipes had corroded and had been leaking the chemical into the ground. After stopping the leak, the company sank observation and sampling wells and found that the product was sitting in a vertical plume, slowly diffusing into a deep aquifer. Because there was no surface or groundwater pollution off the plant property, the plant manager decided to do nothing. Jesse thought that somewhere under the plant there still sits this plume, although the last tests from the sampling wells showed that the concentration of the chemical in the groundwater within 400 feet of the surface was essentially zero. The wells were capped, and the story never appeared in the press.

Peter is taken aback by this apparently innocent revelation. He recognizes that state law requires him to report all spills, but what about spills that occurred years ago, where the effects of the spill seem to have dissipated? He frowns and says to Jesse, "We have to report this spill to the state, you know." Jesse is incredulous. "But there is no spill. If the state made us look for it, we probably could not find it; and even if we did, it makes no sense whatsoever to pump it out or contain it in any way."

"But the law says that we have to report...," replies Peter.

"Hey, look. I told you this in confidence. Your own engineering code of ethics requires client confidentiality. And what would be the good of going to the state? There is nothing to be done. The only thing that would happen is that the company would get into trouble and have to spend useless dollars to correct a situation that cannot be corrected and does not need remediation."

Peter knows professionals at the company that have worked there long enough and may be able to provide first hand knowledge of what occurred back then. And Peter knows that if a spill of that sort occurred today, there are tests and sampling he'd recommend that the company is not currently conducting. At the same time, Peter is fairly certain that no adverse affects to the surrounding soil would be detected.

Given the free reign that Jesse has provided him, Peter feels he could conduct the extra testing without Jesse knowing about it before the results were available. In any case, Peter would hate for this company and its current senior staff – that is doing such a great job with environmental safety – to pay the price for mistakes made decades ago.

What is the ethical dilemma?
What are the first 3 steps the engineer should take?

IMPORTANCE OF ETHICAL KNOW-HOW

How simple or straight-forward was it to define the ethical dilemma? How easy was it to identify what Peter should do? Should he try to determine on his own if there are any detrimental effects from the spill? Should he report the spill no matter the consequences to his own career, the company's reputation, and the company's profit?

How would you decide this? What ethical principles can you look at to guide you through decisions like these?

These are the questions this book will explore. Ethical dilemmas are often not easy to describe concisely and clearly. The steps to take to address ethical dilemmas are often unclear.

It's important to have a means for defining and describing ethical dilemmas. It's important to know what ethical guiding principles you are expected to uphold. It's also important to have a strong sense of your own personal ethical values and priorities.

These are important so that you'll be able to bring clarity to a muddy situation and be able to act in ways that you know are "right" and aligned with stated ethical guidelines and what you believe to be most important.

In this situation the first steps Peter should involve gathering the facts. There are a number of steps he may take to do that such as reviewing paperwork or findings about the leak, obtaining copies of testing and sampling documents from that time period, talking to company employees who worked at the company at that time, and/or checking the law regarding the reporting of historical leaks.

- It can be difficult to clearly define or describe ethical issues
- The "right" action to take is often unclear
- It's important to:
  — Have a means for defining and describing ethical dilemmas

— Know what ethical principles you are expected to uphold
— Have a strong sense of your personal ethical values and priorities

LEARNING OBJECTIVES

By the end of this section, you will:

• Have a picture of what ethical behavior looks like
• Have ideas for what to do if you or someone else is not behaving in a manner consistent with this picture
• Be able to use steps to solve ethical dilemmas

Given these learning objectives, what would you personally like to achieve upon completion of this section? Write down 1 or 2 personal learning objectives for yourself. Examples might be something like: Be able to write my personal ethical guidelines OR Thoroughly understand my company's code of ethics OR Know what to do in a specific situation that you're currently addressing.

Write 1 or 2 personal learning objectives before reading further.

What objectives would you personally like to achieve upon completion of this section of the book?

PART 1 OUTLINE

Part 1 is outlined as follows:

You're currently reading **Chapter 1 – The History of Engineering Ethics**. This chapter is designed to give you a good feel for the history of engineering ethics – why and how they originated and how they have evolved over the years.

Chapter 2 is The Shape of Engineering Ethics. This chapter will familiarize you with online resources that provide the latest thought on engineering ethics and have you exploring what affects a person's ethical values.

Chapter 3 is Ethical Concerns. This chapter will prompt you to consider engineering-specific and company-specific ethical considerations.

Chapter 4 is Factors Influencing Ethical Values. This chapter explores the factors that influence your ethical values both inside and outside of the workplace.

Chapter 5 is Solving Ethical Problems. This chapter presents a step-by-step process for solving ethical problems. You will consider how to apply each of the steps in the process.

WHAT HAS HELPED SHAPE ENGINEERING ETHICS?

Which of the following events and phenomena have helped shape engineering ethics?

- Tay Bridge collapse in 1879
- Growing professionalism in the nineteenth century
- Requirement for formal credentials in order to practice
- Boston molasses disaster
- Quebec Bridge collapse in 1907

- Broughton Suspension bridge in England collapses due to faulty design
- Evolution from a personal to professional concern
- Establishment of engineering societies

THE HISTORY OF ENGINEERING ETHICS

Here is a brief synopsis of the history of engineering ethics.

In the nineteenth century a number of phenomena occurred. Concern about public safety grew, especially after the Tay Bridge collapsed in 1879, killing at least 60 people. At the same time, engineering rose as a distinct profession. Engineers, at the time, were either independent professional practitioners or technical employees of large enterprises. There was tension between these two camps as large enterprises fought to maintain control of their employees. Engineering societies then began forming. ASME, for example, was founded in 1880. Even though 4 different engineering societies were formed around this time, ethics was still viewed as simply a personal concern and not as a broad professional concern.

In the nineteenth century...

- Concern about public safety grew
- The Tay Bridge collapsed in 1879, killing at least 60 people

9

- Engineering rose as a distinct profession
- Engineers were either independent professional practitioners or technical employees of large enterprises
- There was tension between these two camps as large enterprises fought to maintain control of their employees
- Engineering societies began forming
- Ethics was viewed as a personal concern and not as a broad professional concern

Around the turn of the twentieth century was a turning point in the history of engineering ethics. There were a series of significant structural failures about this time, including a number of hugely impactful bridge failures. These had a profound effect on engineers and forced the profession as a whole to confront shortcomings in technical and construction practice and, as a result, ethical standards.

A few things began to happen. Entities began to require formal credentials or licensing involving some combination of educational, experience, and testing requirements. Most engineering societies adopted formal codes of ethics. ASME did so in 1914. We'll explore these codes of ethics in more detail later in the chapter.

Around the turn of the twentieth century:

- There were a series of significant structural failures, including a number of impactful bridge failures
- These forced the profession as a whole to confront shortcomings in technical and construction practice and, as a result, ethical standards
- Entities began to require formal credentials or licensing involving some combination of educational, experience, and testing requirements
- Most engineering societies adopted formal codes of ethics

The US model has generally been only to require those practicing independently, i.e. consulting engineers, to be licensed, while engineers working in industry, education, and sometimes government need not be licensed. This has perpetuated the split between professional engineers and those in industry. Professional societies have adopted generally uniform codes of ethics. On the other hand technical societies have generally not adopted these, but instead sometimes offer ethics education and resources to members similar to those of the professional societies. This is not uniform, and the question of who is to be held in the highest regard: the public or the employer, is still an open one in industry, and sometimes in professional practice.

For more information about the history of engineering ethics – including links to data about each of the significant bridge failures – see Wikipedia's section on engineering ethics.

The US model:

- Generally requires those practicing independently to be licensed, while engineers working in industry, education, and sometimes government need not be licensed

- This has perpetuated the split between professional engineers and those in industry
- Professional societies have adopted generally uniform codes of ethics
- Technical societies often offer ethics education and resources to members similar to those of the professional societies
- The question of who is to be held in the highest regard: the public or the employer, is still an open one in industry, and sometimes in professional practice

WHAT DO YOU THINK OR RECALL?

What do you think or recall is covered in the Code of Conduct for Engineers, originated in 1946?

Consider in your own words what you recall is covered or what you think is covered in the Code of Conduct for Engineers.

Engineers shall...what?

THE CODE OF CONDUCT FOR ENGINEERS

In 1946, the board of directors for the National Society of Professional Engineers – NSPE – approved a "Canons of Ethics for Engineers." Since then, a number of changes and amendments have resulted in today's "Code of Conduct for Engineers."

These canons are stated fairly simply, yet they are all encompassing. There are ways in which you are responsible to the public – in word and deed, to yourself, to your employers and clients, and to your profession.

Take a few moments to navigate to the NSPE website and review the rules of practice that support these canons.

The fundamental canons of the Code of Conduct for Engineers state that engineers shall:

- Hold paramount the safety, health, and welfare of the public
- Perform services only in areas of their competence
- Issue public statements only in an objective and truthful manner
- Act for each employer or client as faithful agents or trustees
- Avoid deceptive acts
- Conduct themselves honorably, responsibly, ethically, and lawfully so as to enhance the honor, reputation, and usefulness of the profession

The NSPE website (http://www.nspe.org/Ethics/CodeofEthics/index.html) outlines all of the rules of practice that support these canons.

THE SHAPE OF ENGINEERING ETHICS

INTRODUCTION

This chapter will familiarize you with online resources that provide the latest thought on engineering ethics and have you exploring what affects a person's ethical values.

What does the research show?

Research online the most recent ethical guidelines for mechanical or professional engineers.

Make note of the websites you feel provide the most relevant or up-to-date guidelines. You may already be familiar with some websites. If you are, navigate to those sites and review the ethical information presented.

Search out the sites that you feel provide the most relevant and up-to-date guidelines.

CURRENT ENGINEERING ETHICS

A few online resources:

Online Ethics Center for Engineering and Research provides a collection of codes of ethics from various professional societies
http://www.onlineethics.org/Resources/ethcodes/21733.aspx

NSPE Code of Ethics for Engineers
http://www.nspe.org/Ethics/CodeofEthics/index.html

ASME Code of Ethics of Engineers
http://files.asme.org/ASMEORG/Governance/3675.pdf

AIChE Code of Ethics
http://www.aiche.org/About/Code.aspx

IMechE
http://www.imeche.org/membership/ethics/

Engineers Australia
http://www.engineersaustralia.org.au//sites/default/files/shado/About%20Us/Overview/Governance/codeofethics2010.pdf

Other Online Resources

Here are a few of the online resources that demonstrate the latest thoughts on engineering ethics. You may have found others as well.

- The Online Ethics Center for Engineering and Research provides a collection of codes of ethics from various professional societies. There are links to the Codes of Ethics published by a number of professional societies including IEEE, the Institute of Electrical and Electronics Engineers; ASCE, the American Society of Civil Engineers; ACM, the Association for Computing Machinery; ASME; and APS, the American Physical Society.
- NSPE – as we already saw – presents a Code of Ethics for Engineers with rules of practice on its website. NSPE is an American engineering society advocating a code of ethics across the engineering field regardless of practice area or employment sector.
- ASME maintains its own Code of Ethics of Engineers as one of its Society Policies. This policy expands upon what was created by NSPE.

This is by no means a comprehensive list, but it will give you a feel for the latest thoughts on engineering ethics. You'll find common language in many of the Codes of Ethics published by these various organizations. Some go into much more detail than others. Some cover guidelines that would apply to all professionals and not just those in engineering. "Never tolerate harassment" being an example of this. Consider which code best applies to you, given the type of work you do.

Your Ethical Values

Look for yourself. What has affected your ethical values? Write down the people, the situations, the beliefs you've been exposed to, the ethical teachings you've been exposed to – anything that has influenced your ethical values in some way.

Take a few minutes to note these influences.

Potential Influences on Our Ethical Values

Here is a list of potential influences on our ethical values.

- Legislation and governing bodies – As new legislation is passed or governing bodies begin to audit or enforce compliance in new ways, our attention is directed in new ways. There are a variety of governing bodies – as you can see form the small sampling of logos on this slide – that work to protect us in the areas of safe work practices, environmental protection, adherence to labor laws, compliance with ethical guidelines and more.
- Perceived pressure – The type and level of pressure we experience influences us. Who is pressuring us? What are the consequences of not fulfilling on the company's, the client's, the community's, our manager's, or our own expectations?
- Company culture – The cultures of your current and past companies impact your ethical values for a number of reasons.

— Policies and practices – Do the stated policies and practices encourage or seemingly discourage ethical behavior?

— Negotiation tactics – In particular the negotiation practices employed by a company can influence one's ethics. Is it a common practice to "pad" your numbers as you're initially bidding projects? Is it a common practice to underestimate your sales forecasts so you can exceed expectations?

— Reward systems – What do the company's reward systems actually reward? Are employees encouraged to get the projects done on time no matter what? Are high sales values rewarded, regardless of the consequences to production or to the clients?

— Code of conduct – Does the company have a formal code of conduct? Does the company train employees regularly on the use and application of the code of conduct?

All of these factors and more will influence a company's ethical culture.

Potential Influences on Our Ethical Values

- Legislation and governing bodies
- Perceived pressure
- Company culture
  - Policies and practices
  - Negotiation tactics
  - Reward systems
  - Code of conduct

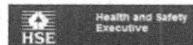

There are more potential influences on our ethical values:

- Media – As greater media attention is given to unethical acts, people might more closely consider their actions. The variety and amount of media that influences us is ever increasing – as you can see by the small sampling of logos on the slide.
- Personal experience – Personal experience obviously influences your ethical values. Have you witnessed double standards of behavior? Have you encountered management that doesn't fulfill on their promises? Also the knowledge you have in technical fields and other areas impacts how you view a situation. And all the other influences you identified on the previous slide influence your ethical principles.
- Personal style preferences – Some of us innately prefer to follow the rules no matter what and some of us are more prone to break the rules when we think the circumstances warrant it. Note, one natural tendency does not necessarily make us more ethical than the other. Sometimes, the most ethical thing to do is to break a stated rule. Other times the most ethical thing to do is to adhere closely to all stated rules.
- Our position – Our level of authority and our perceived level of authority impacts our ethical values.
- Opportunity – What is the perceived opportunity to take advantage of a situation? There's a 10-40-40-10 rule that has been studied and associated with perceived opportunity. It states that 10% will take advantage of a situation if the risk is low. 40% will go along with the work group. 40% will try to determine what the organization wants them to do and comply. And 10% feel the organization has not gone far enough and push for their own personal beliefs.

Potential Influences on Our Ethical Values

- Media
- Personal experience
- Personal style preferences
- Our position
- Opportunity

Google

NEW YORK POST

BBC

abc

Chicago Tribune

This concludes Chapter 2: The Shape of Engineering Ethics.

CHAPTER 3

ETHICAL CONCERNS

INTRODUCTION

This chapter will prompt you to consider engineering-specific and company-specific ethical considerations.

Read about the engineering projects below and consider the answers to the questions that follow.

Project 1:

What about this project?

Your company is designing a new type of front air suspension for on-highway trucks. The design process is being done in partnership with a key truck customer. The agreement between the companies states an aggressive launch date of March 1. First year build rate is 1,000 suspensions with a projected 5% increase for the next 3 years.

Questions to consider:

- How many people are affected by this project?
- What pressures might an engineer encounter during the design and development processes involved in this project?
- What types of obstacles and potential dangers are inherent in this project?

Project 2:

What about this project?

A neighboring town with a population of 8,200 has asked your company about the viability and resources needed to build a drawbridge over the south branch of the river. The road on either side of the bridge site has been built and closed off for years. The town council would like your findings in 3 months' time.

Questions to consider:

- How many people are affected by this project?
- What pressures might an engineer encounter during the design and development processes involved in this project?
- What types of obstacles and potential dangers are inherent in this project?

ETHICAL CONCERNS FOR ENGINEERS

Given what you thought about the projects above and what you know about the nature of your work, why is it important for engineers – in particular – to be concerned with ethics?

Take a few moments to consider your thoughts.

Here are reasons why ethics are of a special concern to the engineering profession:

- Engineering is an important and learned profession.
- Engineering projects can directly and indirectly affect people's quality of life.
- The role of engineer takes different formats: There are professional engineers who are licensed or registered, who often enjoy significant influence over their regulation. There are engineers in industry who have more of an employee-employer relationship, who are influenced by organizational codes of conduct. There are engineers in a consultative role, who might have a slightly different set of principles for working with clients.
- There is a high likelihood of competing priorities – such as controlling costs, minimizing risk to the public, and meeting expected timelines – in engineering projects.

- The necessity for various disciplines and/or entities to collaborate on engineering projects is high, and this increases the possibility for ethical issues such as conflicts of interest, inappropriate treatment of confidential information, kickbacks, differing stakeholder interests, etc.
- Engineering projects can be resource heavy. There can be a significant amount of financial, environmental, and human resources involved.

Ethical Concerns for Engineers

Rationale as to why ethics are of a special concern to the engineering profession:

- Engineering is an important and learned profession
- Engineering projects can directly and indirectly affect people's quality of life
- The role of engineer takes different formats:
  - Professional engineers who are licensed or registered
  - Engineers in industry who have more of an employee-employer relationship
  - Engineers in a consultative role

- The likelihood of competing priorities in engineering projects (controlling costs, minimizing risk to the public, meeting expected timelines, etc.)
- The necessity for various disciplines and/or entities to collaborate on engineering projects (increases possibility for conflicts of interest, inappropriate treatment of confidential information, kickbacks, differing stakeholder interests, etc.)
- Engineering projects can be resource heavy

For these reasons acting ethically is of a particular concern to the engineering field.

Why might a company adopt a code of ethics?

Time yourself for 5 minutes. In that time, identify as many reasons as you can as to why a company might adopt a code of ethics.

Consider:

— What are the benefits to the organization itself?
— What are the benefits to its employees?
— What are the benefits to its community?

Consider as many reasons as you can in 5 minutes.

COMPANY CODE OF ETHICS

Why do companies adopt a code of ethics?

Here are reasons why a company might adopt a code of ethics. Read through them and see if you were able to identify all of these reasons. Perhaps you were able to identify more specific reasons.

Pause for a moment as you review why a company might adopt a business code of ethics.

Company Code of Ethics

Why do companies adopt a code of ethics?

To be a highly ethical organization

To protect the organization against litigation

To have the reputation as a highly ethical organization

To protect company assets

To provide employees with guidance about how to avoid ethical problems

To inform employees how to report violations

To educate employees about legal and illegal activities

To help keep employees safe

To help keep customers and the community safe

When we know why a company might adopt a code of ethics, we can begin to imagine what might be covered in a code of ethics.

What might a company code of ethics cover?

Here is a list of topics that might be covered in a company's code of ethics. Take a moment to review the list.

What do you feel it would be most important for a company to address in their code of ethics?

Company Code of Ethics

What might a company code of ethics cover?

Administering and enforcing the code of ethics

Antitrust laws, restrictive trade practices, boycotts, and technology transfer restriction

Accuracy and retention of business records

Gifts and entertainment

Conflicts of interest

Political contributions and lobbying activities

Protecting company property and information

Securities laws

Business relationships with non-company individuals and entities

Health, environmental, and safety protection

Workplace practices (anti-harassment, anti-violence, drug testing)

See the links below the following logos to see the codes of ethics that these organizations have created and published.

Company Codes of Ethics

GE Citizenship

Rolls-Royce

bp

Westinghouse

BOEING

http://www.rolls-royce.com/sustainability/better_business/ethics/

http://www.wecnq.com/docs/code_of_ethics.pdf

http://www.bp.com/en/global/corporate/about-bp/company-information/code-of-conduct.html

http://www.ge.com/en/citizenship/govcomp-jun20/spirit/

http://www.boeing.com/boeing/companyoffices/aboutus/ethics/index.page?

This concludes Chapter 3: Ethical Concerns.

FACTORS INFLUENCING ETHICAL VALUES

INTRODUCTION

This chapter explores the factors that influence your ethical values both inside and outside of the workplace.

Your Engineering Projects

List of all the engineering projects in which you are currently involved.

For each project create a chart listing the project, the people affected by the end design, the potential safety and environmental risks and impacts, resources and capital involved, your responsibility on the project, and your ethical responsibility.

   Create a chart like the one shown that includes information for all the engineering projects in which you are currently involved. Do this before advancing to the next slide.

Project	People affected	Potential safety & environmental risks & impacts	Resources & capital involved	Your responsibility	Your ethical responsibility

Engineering Projects

Here is a sample completed chart.

For your chart, make note of the impact and the significance of the impact that you have and that the overall project has. For some projects, this can be sizable. The projects you work on affect people, places, and dollars spent. There is a significant responsibility that comes with this.

   You have an ethical responsibility to yourself, to your co-workers, to your company, to your customers, and to the public.

Project	People affected	Potential safety & environmental risks & impacts	Resources & capital involved	Your responsibility	Your ethical responsibility
Replacing a formed and machined metal component in a refrigerator with a molded engineered plastic product	Production employees; End customers	Low risk of fire hazard; Positive impacts for environment	Change to assembly flow; R&D; Renders equipment obsolete; Purchase of injection molding machine; 130K-180K	Lead engineer; Design that meets reqs from a primary distributor, operations, and purchasing; Meet timeline; Stay under budget	Maximize positive environment impacts; Minimize ergonomic issues for production; Eliminate any hazards for the end customer; Make good use of company $$ & time

ETHICAL ISSUES OUTSIDE OF WORK

Rhonda has recently encountered some circumstances at her child's school and at her condominium association that may involve ethical issues. Consider each set of circumstances and then answer the questions: What ethical issue is Rhonda faced with in this scenario? What ethical guidelines would you apply in this situation?

Let's start with the circumstances at her child's school.

Rhonda is attending her son's football game. The stands are pretty crowded with parents and students. She can hear one of her co-workers, John, a few rows back. She's about to turn around and say "Hello," when she realizes he is in the middle of a conversation. The other person asks John, "What are you working on these days?" John replies, "I just started a new exciting project. It's confidential, but we just started collecting bids to expand the water treatment plant at the south end of town. I would get the opportunity to manage the budget for this one – so it would mean that I'm moving up in the organization!"

- What ethical issue is Rhonda faced with?
- What ethical guidelines would you apply in this situation?

At her child's school

In her condominium association

Next let's consider the circumstances in her condo association board.

Rhonda serves on her condominium's association board. The board is responsible for maintaining all interior common areas, the exterior of the building, and the property. Rhonda's condo building has 112 units and a relatively sizable budget to manage. The board recently voted to start working with a new landscaping company, AB Seed. Frank, the Treasurer for the condo board, suggested this company because his sister runs it and he was able to talk her into giving the condo association a deep discount for services. Everyone felt it was a great deal. After the committee passed the vote to start work with AB Seed, Rhonda learned that Frank is part owner in the company. He seems to be a silent partner, yet he has a financial interest in this company.

- What ethical issue is Rhonda faced with?
- What ethical guidelines would you apply in this situation?

Consider these questions regarding the circumstances at Rhonda's condominium association.

What potential ethical issues is Rhonda facing outside of the workplace?

- Potential inappropriate treatment of confidential information as she overhears a co-worker at her kid's school
- Potential conflict of interest in her condo association board when the person recommending a supplier has a financial interest in said supplier

At her child's school
inappropriate treatment of confidential information

In her condominium association
conflict of interest

ETHICAL ISSUES IN THE WORKPLACE

How Can Ethical Values Change?

Andy had been certain that whenever there was a significant potential harmful impact on the environment, he'd share his views with the appropriate authority and refuse to be involved.

Now he's considering heading up a project that will result in routine releases of chemical hazards into his community's environment.

Andy has felt that he has always had a strong set of personal ethical guiding principles. These have allowed him to objectively assess a situation and make the choice that he feels is most aligned with his ethical values.

Recently, his ethical guiding principles have been challenged. They don't seem to be as clear cut as they once were.

There is a project that Andy's company is bidding on. One that Andy feels should not be completed. He doesn't feel it's in the best interest of the community to build another manufacturing plant that will have routine releases of hazardous chemicals into the environment. The community already has a number of manufacturing operations that do this. Andy is surprised the local government has welcomed this new manufacturing facility to town.

His Director wants to assign this project to Andy's team, should their organization be awarded the business. Andy doesn't feel good about having himself and his team involved in designing this manufacturing facility.

Yet, Andy is considering setting aside his concerns for the community because of other information he has recently learned. Andy recently overheard a conversation between his Director and another staff member and learned that the work his team completes is being considered for outsourcing.

Andy feels that his job would be safe if the outsourcing took place. Yet, he knows most – if not all – of his team members would be without a job. In their small community, this would be financially difficult for many of them. One has a new baby, another has 3 kids in college, and the spouse of another just got diagnosed with cancer. Andy doesn't want to think about telling these folks that they no longer have a job.

No one knows that Andy overheard the conversation about outsourcing. Normally, Andy would not take into account information that he did not supposedly have or didn't have the chance to verify.

If his team doesn't complete the project, another team will. There are a number of organizations bidding on the project, and Andy knows one of them will be awarded the work. His act of saying "No" to the project will not stop the project from moving forward.

Previous to this set of circumstances, Andy would've said "No" to this project. Now he's considering having his team complete the project.

What would you say has modified Andy's beliefs? And why?

Take a few minutes and write down your thoughts.

Common Factors that Influence Ethical Values

- Motives of persons involved
- Perception of control
- Perceived alternatives
- Perceived consequences and probability of consequences
- Personal experience

Here are common factors that can influence ethical values:

- Motives of persons involved – What do you care most about? What do the people around you care most about?
- Perception of control – How certain are you that you can control the outcome of the situation? Who do you think does control the outcome?
- Perceived alternatives – How many alternatives are there? How attractive are the viable alternatives? What is the potential impact of each of the alternatives?
- Perceived consequences and probability of consequences – What is the likelihood of experiencing a negative impact – either for yourself or for others? How serious would the consequences be?
- Personal experience – What kind of exposure have you had to highly ethical or unethical behavior? How have you seen others stick to their values no matter what? How have you seen others bend the rules? What's been your experience when you've acted with integrity or when you've bent the rules?

How does this apply to you?

For each of the <u>factors influencing ethical values,</u> think of a situation where the factor might have an influence on your ethical values.

- •motives of persons involved
- •perception of control
- •perceived alternatives
- •perceived consequences and probability of consequences
- •personal experience

How might these factors influence you? For each of the factors that might influence ethical values, provide an example of a situation in which the factor might influence your ethical values.

Identify 5 situations – one for each factor. You might be able to use the same situation more than once.

It's valuable to be aware of the ethical factors at play in the work that you do.

This concludes Chapter 4: Factors Influencing Ethical Values.

SOLVING ETHICAL PROBLEMS

CHAPTER

This chapter presents a step-by-step process for solving ethical problems. You will consider how to apply each of the steps in the process.

What Would You Do?

Read the situation described below then answer the question: How would you solve this ethical issue? Think about the situation and then write out the specific steps and approach you'd use to solve the issue at hand.

What would you do?

You're an engineer on a new product development team. You were surprised by the choice the Director of Purchasing made for the ball bearing supplier. The word around the water cooler is that this supplier – Bearings Unlimited – is owned by your CEO's brother-in-law. You know the Director of Procurement to be a great guy and you can't imagine him not making a fair choice about a supplier. You had almost fully set your concerns aside when you had a conversation with another supplier and your contact casually mentions that they haven't gotten any new RFQs from your organization in over a year. The procurement plans for this new product included sending an RFQ to this supplier. Why did that not happen? Did it happen and your contact didn't know about it? Did the RFQ get lost between here and there? You sense something fishy is going on.

Do you have a current project that involves ethical issues or you think might involve ethical issues? If you do, take a few moments and write down what you think are the most appropriate next steps for you to take to solve the issue.

Process for Solving Ethical Problems

Outlined here is an 8-step process for solving ethical issues.

Step 1 – GATHER the facts. Ethical dilemmas often start out with hearsay and hunches and gut feelings. That's normal, and you want to find out what objectively are the facts.

Step 2 – DETERMINE the values and ethical issues involved. Ethical dilemmas often stem from competing ethical principles. What seems to be the best thing to do conflicts with governmental rules OR doing the best thing for one group means potentially harming another. Even if ethical values do not seem to be in direct combat with one another, it's useful to outline all the ethical values at play.

Step 3 – DECIDE whether you believe unethical behavior has occurred or is occurring. Once you are armed with the facts and have objectively outlined the ethical values involved, you'll be in a better position to decide whether you believe unethical behavior is taking place. And notice that you don't decide this until step 3 in the process. This keeps us from jumping to conclusions too quickly.

Step 4 – RANK in prioritized order the ethical principles involved. Ranking the principles may feel quite straightforward in some situations and more subjective in others. Having this prioritized order will help you – and the others involved – develop effective action plans.

Step 5 – REPORT the unethical behavior. Informally to any persons who need to be informed in order to do their jobs and formally to any governing or oversight entities.

Step 6 – DEVELOP an action plan to address the unethical behavior. In many cases this will be a group effort. And, in some circumstances you may have to complete this on your own.

Step 7 – IMPLEMENT the action plan. Make sure the tasks are done in the agreed-upon or stated time frames.

Step 8 – REFLECT on the outcome. Was the unethical behavior stopped or resolved? What did you learn from the process? How do you know you did the quote-unquote "right" thing?

Let's look at the process in more detail.

Process for Solving Ethical Problems

1. GATHER the facts
2. DETERMINE the values and ethical issues involved
3. DECIDE whether you believe unethical behavior has occurred or is occurring
4. RANK in prioritized order the ethical principles involved
5. REPORT the unethical behavior
6. DEVELOP an action plan to address the unethical behavior
7. IMPLMENT the action plan
8. REFLECT on the outcome

There are some things you know such as which suppliers were to have received the RFQ for this project.

There are some things you don't know such as whether the selected supplier's owner is related to your CEO, the selection criteria used, and whether all the suppliers that were to have received RFQ did in fact receive RFQs.

Given this, how would you go about gathering the facts?

What specific actions will you take? Outline your approach.

If you have a current project that potentially involves ethical issues, consider the same question for that project. How will you go about gathering the facts?

> # 1. GATHER the facts
>
> - You haven't been directly involved in the supplier selection process for the ball bearing component of the new product
> - You question whether a supplier – that was supposed to receive an RFQ – was intentionally left out of the RFQ process
> - You've heard the owner of the supplier selected is related to your CEO
> - You don't know the selection criteria used for selection of this supplier

Let's say you've gathered your facts. You've spoken to the procurement professional who sends out the RFQs and she had no record of RFQs being sent to 2 of the suppliers that were to have received them.

You ask the other people on the new product development team if a new decision had been made about whom was to receive RFQs. No one is aware of any change to the original choices.

As hard as it was, you found a time to ask your CEO if he is indeed the brother-in-law of Bearings Unlimited's owner, like you had heard. He confirmed that he is and added, "I hope he wasn't given any special treatment because of that."

You ask to see the supplier selection criteria used. You say you're curious about why a certain supplier didn't measure up. And, sure enough, the 2 suppliers in question do not show up on the comparison chart.

You ask the Director of Procurement why the 2 suppliers don't show up on the supplier selection criteria chart. He says, "They must not have responded to the RFQ." You say that Becky didn't have any record of them being sent RFQs. He responds, "Well, it's too late now. It would cost too much money now to back out and go another direction." You know this. You also know that the project is already behind schedule. You then swallow hard and ask, "Were these 2 suppliers deliberately left out of the process? And does our CEO being related to Bearings Unlimited's owner have anything to do with this?" He responds, "I'm shocked and hurt you would even ask me that. The answers to your questions are 'No'."

With these facts in hand, what values and ethical issues do you believe are involved?

If you have a current project that potentially involves ethical issues, consider the same question for that project. Given the facts that you currently have, what values and ethical issues do you believe are involved? You may not yet have enough facts to be certain. Write down what you can surmise at this point.

---

2. DETERMINE the values and ethical issues involved

- There is no record of RFQs being sent to 2 of the 5 suppliers that were to have received them
- The CEO is the brother-in-law of Bearings Unlimited's owner
- Only 3 suppliers show up on the supplier selection criteria comparison chart
- The Director of Procurement says it's too late, it would cost unnecessary money to reevaluate the suppliers, and there wasn't any intentional wrong-doing
- The project is currently behind schedule

You may have identified ethical issues similar to the following:

- A potential conflict of interest – There seems to be a question of whether the CEO's relationship had an inappropriate bearing on the choice made.
- In any case, there is a perception of a conflict of interest.
- Another potential ethical issue in this situation is deception – not including 2 suppliers the project team decided should be included.

- There is also a product quality responsibility to the employer and the customer. The company might produce a product with a higher level of quality by using of one of these 2 suppliers.
- And there is a financial responsibility to both the employer and the customer. Changes to the supplier used would cost more in the short-term and delays to the project would have PR and financial repercussions to employer and to the customer.

See how all the nuances of the situation come to light as you outline the different ethical values and issues at play? You begin to see the ripple effects of the actions taken and the actions that could be taken.

Given the ethical issues that are involved, what would you decide? Do you believe unethical behavior has occurred or is occurring? Consider and make your decision.

If you have a current project that potentially involves ethical issues, consider the same question for that project. Do you believe unethical behavior has occurred or is occurring?

3. DECIDE whether you believe unethical behavior has occurred or is occurring

- Conflict of interest – Did CEO's relationship have an inappropriate bearing on the choice made?
- Perception of a conflict of interest
- Deception – not including 2 suppliers the project team decided should be included
- Product quality responsibility to employer/customer – product might be of a higher quality through use of one of these 2 suppliers
- Financial responsibility to employer/customer – changes to the supplier would cost more in the short-term and delays to the project would have PR and financial repercussions to employer and to the customer

Let's say that yes, you do believe that unethical behavior has occurred and is continuing to occur because no one is looking at reconsidering the choices made to date. Whether intentional or not, you believe those 2 suppliers should not have been left out of the RFQ process and you are concerned with the implications of that.

Your choice and reasoning might have differed from this. For the sake of moving forward with the example, please assume the same decision.

Given that choice, how would you rank in prioritized order the ethical principles involved. Which ethical issues are most prevalent? Which ethical principles are most important to the situation at hand? How relevant and how important they are must be weighed as you're prioritizing. There is not a perfect formula for this. You need to prioritize the ethical principles involved.

If you have a current project that potentially involves ethical issues, consider the same question for that project. Given an assumed decision that yes unethical behavior has occurred or is occurring, how would you rank in prioritized order the ethical principles involved in your current project?

---

4. RANK in prioritized order the ethical principles involved

- Yes, you do believe that unethical behavior has occurred and is continuing to occur because no one is looking at reconsidering the choices made to date
- Whether intentional or not, you believe those 2 suppliers should not have been left out of the RFQ process
- You are concerned with the implications to your employer and to the customer

---

Here is a potential prioritized order for the ethical issues involved. Given your personal experience and other factors that influence our ethical values, your prioritized order may look differently. An important part of this process is getting your reasoning and rationale clarified for yourself so that you can share it with others.

1. Let's say you first prioritized the issue of having a perceived conflict of interest. Whether actions were deliberate or not, it doesn't look good.
2. Next is the concern about deception. If one member of the team deceived the others on this project, it is going to severely damage the trust in the group. And this damaged trust could impact all the work the groups do together in the future. This has negative implications for the employer, the employees involved, the suppliers, and customers on this project and future projects.
3. Next is product quality. There's a good probability – the engineer believes – that the 2 suppliers left out of the selection process could have exceeded the quality of the supplier selected. It's not fair for the customer, nor for the company, to produce a product that's not as high of quality as it could be.
4. The engineer followed this up with the concern about an actual conflict of interest.
5. And then the financial impact. The engineer ranked this one last because he's uncertain about the likelihood and extent of the impact. There would be short-term impacts, certainly. Yet, reevaluating and going with a different supplier may improve costs in the long-term.

And there may be other ethical issues at play. These were the primary one's from the engineer's perspective.

Given this prioritized order of ethical principles. How would you go about reporting the unethical behavior? To whom would you report the unethical behavior and how? Pause a moment while you consider that.

If you have a current project that potentially involves ethical issues, consider the same question for that project. How would you go about reporting the unethical behavior? To whom would you report the unethical behavior and how?

---

5. REPORT the unethical behavior

1. Perception of a conflict of interest
2. Deception – not including 2 suppliers the project team decided should be included
3. Product quality responsibility to employer/customer – product might be of a higher quality through use of one of these 2 suppliers
4. Conflict of interest – Did CEO's relationship have an inappropriate bearing on the choice made?
5. Financial responsibility to employer/customer – changes to the supplier would cost more in the short-term and delays to the project would have PR and financial repercussions to employer and to the customer

---

How to report an ethical issue is going to vary depending on the circumstances. If the company for which the engineer works has a Compliance Committee or a Chief Compliance Officer or specific guidance in a code of ethics about how to report violations, the engineer would go to these resources for guidance. If the work of the project team is overseen by a governing or compliance entity, the engineer can contact those organizations. The engineer can also contact different professional agencies for advice on how to best proceed.

Let's say the engineer in this case contacted the Chief Compliance Officer at his company as his employer's code of ethics instructs him to do. The Chief Compliance Officer thanks the engineer for bringing this to his attention. The Officer lets him know that he's going to have conversations with the CEO and the Director of Procurement regarding the situation and then would like the help of the new product development team to co-create the plan for addressing the situation. The Officer will be in touch soon.

Given this, how would you develop an action plan? What actions do you think the team should take to address the unethical behavior? Write out the action steps.

If you have a current project that potentially involves ethical issues, consider the same question for that project. What actions do you think should be taken to address the potential unethical behavior?

---

6. DEVELOP an action plan to address the unethical behavior

- Could report internally to a Compliance Committee member or as a company code of ethics instructs you
- Could report externally to any governing or compliance entities
- Could contact professional associations for advice

Engineer contacted his Corporate Compliance Officer as instructed in his company's code of conduct.

Officer will talk with CEO and Director of Procurement and then would like the project team to meet to action plan.

The dynamic will change with the full project team involved. The members will likely have differing perspectives on the ethical principles involved and their prioritized order. In fact, the first step the team might take is to review and align on the prioritized ethical issues involved.

Let's say the project team decided to take these 3 steps in this order:

1. Align on and prioritize the ethical issues involved.
2. Communicate the prioritized ethical issues to the CEO.
3. Send RFQs to the 2 suppliers who did not yet receive the RFQs.

At this point the team will see if and how the suppliers respond and will then decide what actions to take next. Answer the questions: How would you complete the action plan? How would you ensure the action plan is implemented? How will you ensure the action planning continues?

If you have a current project that potentially involves ethical issues, consider the same question for that project. How will you ensure that the action plan is implemented?

7. IMPLEMENT the action plan

1. Align on and prioritize the ethical issues involved
2. Communicate the prioritized ethical issues to the CEO
3. Send RFQs to the 2 suppliers who did not yet receive the RFQs

To ensure that action plans are implemented the team assigned an owner and a due date for each action. And the team has scheduled regular weekly meetings to assess progress and continue creating plans to move forward. Also, the team has decided to include a representative from the customer in these weekly meetings so that the perspective and knowledge from the customer can be appropriately included in the process.

Given the approach to implementation, how would you reflect on the outcome? How would you reflect on the outcome of the ethical problem solving process? What would you consider? Think about and respond to these questions.

If you have a current project that potentially involves ethical issues, consider the same questions for that project. How would you reflect on the outcome there? What would you reflect on?

8. REFLECT on the outcome

- The team assigned an owner and a due date for each action
- The team has scheduled regular weekly meetings to assess progress and continue creating plans to move forward
- The team invited a representative from the customer to join them in their weekly meetings

You want to refer back to your prioritized list of ethical principles involved and assess whether the principles are being addressed and to what degree. When competing principles are involved, each issue might not be resolved 100%. You want to evaluate whether you're doing all you can and whether all parties affected have been communicated with appropriately.

Ethics for engineers are not always clear-cut and simple. If you follow this 8-step process for solving ethical problems, you can know that you're taking a diligent approach and doing all that you know to do.

Reflecting on the Process

- Refer back to your prioritized list of ethical principles involved and assess whether the principles are being addressed and to what degree
- Know that when competing principles are involved, each issue might not be resolved 100%
- Evaluate whether you're doing all you reasonably can
- Determine whether all parties affected have been communicated with appropriately

Continue to Apply What You've Learned

This is the final section of this chapter. You are nearing the end!

At this point, please identify the 3 most important concepts learned during the training. Reflect and identify these.

Next, write 3 to 5 actions you will take as a result of this training. Actions to take might look like: Review what you feel is the most relevant Code of Conduct for Engineers once a month, Determine whether your company has a code of ethics or process for reporting violations, Write out your personal ethical values, Complete a chart indicating your ethical responsibilities for each of the projects you work on, Apply the ethical problem solving.

This concludes Chapter 5: Solving Ethical Problems.

PART 2:

CHANGING ORGANIZATIONAL CULTURE

EXPLAIN CHANGE AND IDENTIFY THE MOTIVATION BEHIND AN UPCOMING CHANGE

By the end of this section, you will be able to explain "Change;" identify the motivation behind change; create a vision for the future; determine how a change will impact processes and people; identify employee development needs; and write a plan for mobilizing an organization.

To get the most out of this section, you should identify a change that will affect you, your team/department or your organization. As the chapter progresses, consider how to apply the principles to your situation. You should have a pen and paper handy to take notes or try exercises on your own.

Learning Objectives

By the end of this section, you will be able to:

- [ ] Explain "Change"
- [ ] Identify the motivation behind an upcoming change
- [ ] Create a vision for the future culture
- [ ] Determine how a change will impact processes and people
- [ ] Identify employee development needs
- [ ] Write a plan for mobilizing an organization

Effective Change Management enables people to assimilate changes. It connects significant change to the culture of the organization, and allows employees to achieve success throughout the process. After completing this chapter, you will understand what Organizational Change is and how to apply Change Management techniques to your workplace. In addition, you will have a set of tools that will assist you in facilitating successful change within your team or organization.

> ### Purpose of this Section
> "All changes experience resistance; most companies find dealing with that resistance the most difficult part of the project."
> Prosci (2000). Best Practices in Change Management: 254 Companies Share Lessons Learned

In this chapter you'll apply what you learn by identifying the necessary ingredients for a successful change and completing a training action plan. These are the chapters that we'll cover during the course. Remember to keep your pen and paper handy to take notes or to try the exercises.

Part 2 Outline

In this section you will learn how to define "change," identify ways to facilitate positive change, identify common reactions to change, and understand The Change Curve. You will also learn how to develop and implement an assessment.

Explain "Change" and Identify the Motivation Behind and Upcoming Change

Explain "Change" & Identify The Motivation Behind An Upcoming Change

❑ Define "Change"

❑ Identify ways for facilitating positive change

❑ Identify common reactions to change

❑ The Change Curve

❑ How to develop and implement an assessment

A company called Shane Manufacturing, which has used a specific process for making parts for over fifteen years, has made the decision to streamline operations. The organization plans to do this by implementing new manufacturing technology that will change the way workers assemble parts. Shane

Manufacturing has multiple choices for what technology to install and how to implement it. Is this a case for Change Management? How do you define change?

> # Consider . . .
>
> ❑ Shane Manufacturing, which has used a specific process for 15+ years, decided to streamline operations
>
> ❑ Shane plans to implement new manufacturing technology
>
> ❑ It has multiple choices for what to install and its implementation

There are several definitions of change. The word, "Change," which can either be a noun or a verb. In the context of change management, we'll focus on the verb. Let's review a few of them together: the first is to make different or alter; another is to replace by another, or substitute; another meaning is to give and take reciprocally, that is, to exchange; and lastly, to become different, pass from one condition to another. Is your definition close to any of these?

> # Defining Change
>
> To change is to:
>
> ❑ Make different or alter
>
> ❑ Replace by another, substitute
>
> ❑ Give and take reciprocally, to exchange
>
> ❑ Become different, pass from one condition to another

Jeff Hiatt and Timothy J. Creasey theorize that Change Management is the result of the merging of two opposite schools of thought on how to create a continuously improving organization – the engineer's approach and the psychologist's approach. These approaches are outlined next.

Change Management

"... the process, tools and techniques to manage the people-side of business change to achieve the required business outcome, and to realize that business change effectively within the social infrastructure of the workplace."

-Hiatt & Creasey

Change Management: Engineer vs. Psychologist

According to Hiatt and Creasey, "To understand change management as we know it today, you need to consider two converging and predominant fields of thought: an engineer's approach to improving business performance and a psychologist's approach to managing the human-side of change."

Engineers focus on the mechanics of change, while psychologists view how change impacts the employees in the organization.

Combining Different Approaches Equals Change Management

	Engineer	Psychologist
Focus	Processes, systems, structure	People
Business practices	BPR, TQM, ISO 9000, Quality	Human Resources, Organizational Development
Starting point	Business issues or opportunities	Personal change, employee resistance (or potential for resistance)
Measure of success	Business performance, financial and statistical metrics	Job satisfaction, turnover, productivity loss
Perspective on change	*"Shoot the stragglers, carry the wounded"*	*"Help individuals make sense of what the change means to them"*

Source: The Definition and History of Change Management, Hiatt and Creasey http://www.change-management.com/tutorial-definition-history.htm

Alone, neither of these approaches is successful. For successful change, both the engineering and the psychological techniques must be used. When we combine the techniques to facilitate a change, we are utilizing Change Management.

Think about a change you have experienced. How did you know that change was not going smoothly?

Let's go back to Shane Manufacturing to see if the company can effectively streamline its operations without Change Management. The Shane Executive Board decides that it is time to streamline operations by converting to new technology. Shane's CEO, Paul Christy, plays racquetball with the owner of a company that specializes in manufacturing technology. Shane Manufacturing purchases its product and announces to the company, through a memo, that in fourteen months the parts plant will be run exclusively using the new technology.

In the fourteen months prior to the launch, Shane Manufacturing, using almost exclusively a team of consultants, works to ready the plant for the installation of the new technology. Occasionally the new system is mentioned briefly in the company newsletter. Two weeks before the new system is scheduled to be turned on, workers attend training classes. On the day the new system is turned on there is a big banner on the plant entrance that reads, "Welcome to a New Day!"

CHANGE SCENARIO 1: SHANE MANUFACTURING

Change Scenario 1: Shane Manufacturing

❏ Shane Manufacturing announces to employees that it is transforming its manufacturing processes using a new technology

❏ The new process will be implemented within 14 months from the date of the announcement

❏ Shane is using external consultants to prepare the plant for the process and equipment transition

❏ Two weeks before the scheduled launch, employees attend training courses

Let's review the results of Shane's change initiative. In the first month after the new technology is implemented, production dips to 50% of pre-change levels. At the three-month mark, worker turnover has increased and productivity is at 65% of pre-change levels. Twelve months after the new technology is implemented, the Shane Manufacturing parts plant finally reaches pre-change productivity levels, but is still not realizing the increase of productivity. They re-train employees in the hopes that additional training will increase the efficacy of the new system.

What factors made this transition stressful for employees? What factors made this stressful for Shane's leadership?

Change Scenario 1: Shane Manufacturing

❑ Results:

⇨ Month 1: production dips to 50% of pre-change levels

⇨ Month 3: employee turnover increases and productivity reaches 65% of pre-change levels

⇨ Month 12: productivity returns to pre-change levels

❑ What factors made this transition stressful for employees?

❑ What factors made this stressful for Shane's leadership?

Let's review some of the facts, which made this transition stressful for employees. First, the employees did not have any input. Shane used external consultants, to the exclusion of almost all internal personnel, to prepare the plant for installation of the new technology.

The company provided little communication about the new system to employees. Employees did not know how the new system would change the way they work, and lastly, training began only two weeks prior to the employees' transition to the new system. This left them little time to adapt to the new system and comprehend how it would impact their respective jobs.

> ## Change Scenario 1: Shane Manufacturing
>
> ❏ What factors made this transition stressful for employees?
>
> - ⇨ Employees had no input
> - ⇨ Almost exclusive use of external consultants to prepare the plant for installation of the new technology
> - ⇨ Shane provided little communication on the new system
> - ⇨ Employees did not know how the new system would change the way they work
> - ⇨ Training began only two weeks prior to the employees' transition to the new system

What factors made this stressful for Shane's leadership?

The Decision to select the particular supplier was made unilaterally by the CEO, without considering if the supplier could meet Shane's needs. The decision on a new supplier was made for personal reasons (that is, the CEO's friendship with the supplier), not because the supplier offered a quality product for a competitive price.

The production falling dramatically after implementation of new system puts a strain on the company as a whole. Employee turnover increases because employees are disengaged and confused by all of the new changes. High Employee turnover adds a dramatic increase in costs to the company, because the company needs to increase expenditures on recruiting and training new employees.

> ## Change Scenario 1: Shane Manufacturing
>
> ❏ What factors made this stressful for Shane's leadership?
>
> - ⇨ Decision on new supplier was made unilaterally by the CEO
> - ⇨ Decision on new supplier was made for personal reasons, not because this was the best supplier
> - ⇨ Production falls dramatically after implementation of new system, while employee turnover increases

Now, let's take a look at how the company's transitions its processes by implementing Change Management.

The Shane Manufacturing Executive Board decides that it is time to streamline operations by converting to new technology. The Executive Board chooses a team made up of representatives from both corporate and the parts plant to select technology that will best meet the company's needs from several available options.

The technology selection team begins their process by creating a vision of what the organization and specifically, the parts plant, will be once the new technology is implemented. Representatives from all level of staff impacted by the new technology, including an employee from the plant floor participate in the visioning process. As part of the process, Shane Manufacturing customers are interviewed.

After a vision is created and a specific technology is selected and purchased, the CEO of Shane Manufacturing announces the change during the quarterly, all- plant meeting. He discusses why Shane Manufacturing is making the change and lists benefits for individual employees and for the plant as a whole.

Change Scenario 2: Shane Manufacturing

❑ Shane Manufacturing's Executive Board creates a team of representatives from different parts of its business whose goal is to select a technology supplier

❑ All representatives, who would be impacted by the new technology, as well as customers, are interviewed as a part of the *Visioning* process

❑ Shane leadership explains how the change would **benefit** the organization as a whole, as well as its specific impact on employees

---

In the fourteen months prior to go-live, Shane Manufacturing regularly communicates project status using multiple methods of communication, including weekly meetings and newsletters. There is a cautiously optimistic buzz in the plant about the changes that will be coming as project team members bring the technology to the plant floor for "road shows" to demonstrate how it will work.

Two weeks before the new system is scheduled to be turned on, workers begin to attend training classes. On the day the new system is turned on there is a support network in place to assist plant employees with real life use of the system. For one month after the launch day, bi-weekly short refresher training sessions are available for employees that feel they need additional help.

Change Scenario 2: Shane Manufacturing

❑ Shane Manufacturing **communicates** with employees *regularly* in the 14 months prior to the launch about the status of the transition using various media

❑ Shane Manufacturing provides employees with **live demonstrations** of the new technology throughout the period before launch

❑ Employees attend training **before** the launch as well as attend sessions **after** the launch, for reinforcement

---

Let take a look at how the results of this approach differs from the first. In the first month after the new technology is implemented, production dips to 80% of pre-change levels. At the three-month mark, productivity is over 100% of pre-change levels. Twelve months after the new technology is implemented, the Shane Manufacturing parts plant is producing at well over their pre-change levels and is beginning to recoup their investment in the new technology.

Employee surveys show that Shane Manufacturing employees have a high level of job satisfaction. What did Shane Manufacturing do to make this an exciting, positive change for employees?

How does Shane's leadership benefit?

Change Scenario 2: Shane Manufacturing

❏ Results:

    ⇨ 1 month: production dips to 80% of pre-launch levels

    ⇨ 3 months: production exceeds 100% of pre-launch levels

    ⇨ 12 months: production exceeds pre-change levels to such an extent that the company is able to re-coup their investment in the technology

❏ What did Shane Manufacturing do to make this an exciting, positive change for employees?

❏ How does Shane's leadership benefit?

What did Shane Manufacturing do to make this an exciting, positive change for employees?

The company involved all levels of the organization in the process early and often. The process was supported from the top down, including the executive board. The leadership raised awareness and interest early in the project and communicated changes continuously, through the launch date.

The leadership ensured that new technology was demonstrated to employees so that they would understand its impact on their jobs and the organization. Employees received training on new processes and technology as well as support and additional training as needed.

Change Scenario 2: Shane Manufacturing

❏ What did Shane Manufacturing do to make this an exciting, positive change for employees?

⇨ Involved all levels of the organization in the process early and often

⇨ Displayed clear executive sponsorship of the change

⇨ Raised awareness and interest early in the project

⇨ Communicated changes continuously throughout the project

⇨ Showed employees what the new world will look like (demonstrated new technology)

⇨ Provided training on new processes and technology

⇨ Provided support and additional training as needed

How does Shane's leadership benefit? Employees are involved and committed to the change. The workforce is productive and focused. The transition successfully minimized turnover costs, increased retention of well-trained employees, and produced a better return on investment. This in turn increased employee confidence that the change was the right thing to do, which is also referred to as employee buy-in.

In sum, organizations that excel at continuous change do so not only because they employ Change Management techniques, but because they are made up of people who react to change in a positive manner.

Change Scenario 2: Shane Manufacturing

❏ How does Shane's leadership benefit?

⇨ Employees are involved and committed to change

⇨ Productive, focused workforce

⇨ Minimize turnover costs

⇨ Better return on investment (ROI)

⇨ Increased confidence that the change is the right thing to do

---

Some adults react instinctively to change in the same way as they reacted as children: with fear, anticipation, distrust, disgust, resentment, hope, anger, or insecurity. They may feel any mix of emotions upon hearing of the change.

Individuals react to a change based on the level of risk or perceived risk the project or change imposes on them. In their paper, two change management experts note, "The foremost principle of risk communication is that 'perception equals reality'. In other words, what is perceived as real is real in its consequences. Employees react to perceived threats, rather than to 'reality'."

Common perceived threats are related to fear of loss of confidence in ability to perform, decreased power, unforeseen negative impact, and even complete failure in the new environment.

Responding to Change

☐ (Some) adults react to change instinctively; in the same manner as they reacted as children

☐ Individuals may feel any mix of emotions upon hearing of the change

☐ Individuals react to a change based on the level of risk or perceived risk the project or change imposes on them

☐ Common perceived threats are related to fear of loss of confidence in ability to perform, decreased power, unforeseen negative impact, and even complete failure in the new environment.

Even if an individual tends to react positively to change, his perception might change quickly based on "Water Cooler Conversation," both electronically and in person. Employee speculation, on the impact and political ramifications of a change, spreads like wildfire through an organization.

If employees believe that the change will impact them negatively, it becomes a "reality." It is essential for leaders of an organization going through a system transition to communicate facts early and often with employees to stem the flood of gossip and speculation through the company.

The Impact of Water Cooler Conversation

❏ *"Did you hear what they are doing? I heard that after the new system is installed they are going to lay off 25% of staff."*

❏ *"This new process is going to make the entire mailroom staff obsolete."*

❏ *"They don't care about what this does to us. It's all about money in management's pockets."*

There are four stages of change, which are illustrated here, in this change curve. Organizations and individuals move through a series of stages after a changed is announced. The **Y** axis begins at "General" and ends at "Specific." This axis refers to information, which is available to both management and/or the project team. It also refers to the specific information disbursed to employees. The **X** Axis shows the stages of change, which are defined within this model.

As you can see, the more specific the information provided to an employee or project team the higher the commitment to participate in the change process.

The Change Curve

❏ There are four stages of Change

⇨ Organizations and individuals move through a series of stages after a changed is announced.

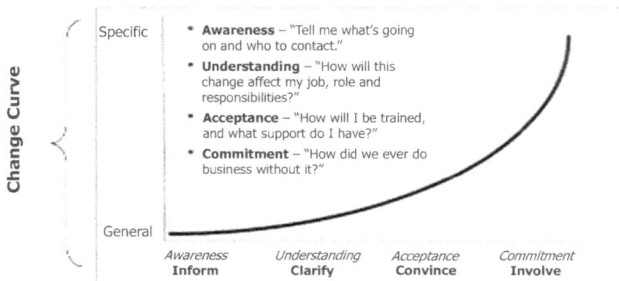

* **Awareness** – "Tell me what's going on and who to contact."
* **Understanding** – "How will this change affect my job, role and responsibilities?"
* **Acceptance** – "How will I be trained, and what support do I have?"
* **Commitment** – "How did we ever do business without it?"

Specific / General — Change Curve

| Awareness | Understanding | Acceptance | Commitment |
| Inform | Clarify | Convince | Involve |

Change management activities are associated with each stage in the process of change. Let's review what each of the six categories means in the context of Change.

Assessment refers to taking a high-level look at the existing organization and how it will be impacted by change.

Vision refers to creation, that is, how leadership views the future after the change has been implemented across the organization.

To **analyze** the impact of change management one needs to identify groups of people affected and how they are affected. One should use the **analysis** to create a change management plan, including areas such as training, communication, and employee support.

There is a great deal of emphasis on the design and implementation of **communications** used to inform employees about the upcoming changes. **Training** refers to the design and implementation of various programs for employees relating to their respective roles in the change process. It is crucial to assesses how **ready** each group is for the implementation of the change. **Support** refers to the activities, which the organization uses to track coaching, reporting data, communicating, tweaking processes, and rewarding successful outcomes.

The Change Curve: Change Management Activities

❏ The six categories of change management activities:

1. Assessment
2. Vision
3. Analysis
4. Communications
5. Readiness
6. Training
7. Support

Think about any changes your department or organization is experiencing, currently. In which stage do you fall on the Change Curve? How about your coworkers? Why?

Reflect...

❏ Think about any changes your department or organization is experiencing, currently

❏ In which stage do you fall on the Change Curve? How about your coworkers? Why?

Try this exercise. Place the change management activities under their corresponding stage of the Change Curve. It is possible that the same activity could occur under more than one stage. Remember, the activities are: assessment, vision, analysis, communications, training, readiness, and support.

Try this:

Place the change management activities under their corresponding stage of the Change Curve. It is possible that the same activity could occur under more than one stage.

Awareness	Understanding	Acceptance	Commitment

Notice that communications is present in each of the Change Curve stages. Ongoing communications not only ensure that the organization has the information it needs, but it can also move through all four stages successfully. Effective communications also minimize the impact of negative "water cooler" conversations among employees.

Change Curve Activities: Answers			
Awareness	**Understanding**	**Acceptance**	**Commitment**
❏ Assessment	❏ Analysis	❏ Training	❏ Communications
❏ Vision Creation	❏ Communications	❏ Readiness	❏ Support
❏ Communications		❏ Communications	
		❏ Support	

At the beginning of a project it is important to spend time taking an in-depth look at the current state of your organization and the goal you are trying to accomplish.

By completing an assessment, you not only gather the essential information you need in order to move to the next step, but you also motivate the rest of your team to think about how the changes are going to impact the organization and its employees.

Assessment

❏ Review the current state of the organization's processes, internal and external communication, performance management, etc.

❏ Completing an assessment allows you to garner critical information

❏ Communicate assessment results to peers and employees

Please review these Questions, which you might ask during an assessment.

At this point in the project, your sources for this information are primarily organizational leadership and the project team (if it has been named). Later in the project, you will gather information from the organization as a whole.

Assessment Questions

❑ Questions you might ask, during an assessment, include:
- ⇨ What is the current situation?
- ⇨ What is the envisioned, future situation?
- ⇨ Why are you making this change?
- ⇨ Who decides if the change will be made?
- ⇨ Who is investing time and/or money in the change?
- ⇨ Who will make decisions along the way?
- ⇨ Who will approve decisions?
- ⇨ Who will need to change their jobs, skis, behaviors, expectations, or processes?
- ⇨ Who will be excited about the change?
- ⇨ Who will not be excited about the change? How will they feel and why?
- ⇨ What communications vehicles are in place in the organization?
- ⇨ What are the training methods customarily used by the organization?
- ⇨ What is the project timeline?

The initiative is called SMART, which is an acronym for, "Shane Manufacturing Automatic parts Routing Transformation. The current situation is that the Shane Manufacturing Executive Board has decided to implement new manufacturing technology at the Cleveland parts plant.

There are multiple vendors that may meet the company's needs, and a specific vendor has not been chosen yet. In the past, large-scale changes have occurred, but not smoothly. In the past, there have been union issues and major blips in production as a result.

The conclusion arrived at after talking with both the Executive Board and team members is that the future situation has not yet really been defined. Most responses were vague and had the general theme of "making things better". Developing a specific vision for the project is a key step.

Sample Change Assessment

❑ **Name of the Initiative**: Shane Manufacturing
Automatic parts Routing Transformation (**SMART**)

1. What is the current situation?

    ★ Cleveland parts plant will be implementing new manufacturing technology

    ★ Multiple vendors from which to chose; determining which meet its needs

    ★ Large-scale changes have not gone smoothly in the past

2. What is the envisioned, future situation?

    ★ The Executive Board and team members have only provided vague answers

        • The future situation is therefore, undefined

There are several reasons why the Executive Board decided that Shane must implement this change. Today, its manual process is not ideal. Over the past five years, customer complaints related to quality and employee injuries have increased.

The Executive Board, the project team, representatives from the plant, and all plant employees are all investing time, money, and resources to support this change and to ensure that it's successful.

Sample Change Assessment (continued)

3. Why are you making this change?

    ★ Over the past five years, customer complaints related to quality and employee injuries have increased.

4. Who decided that this change should be made?

    ★ The Executive board

5. Who is investing time and/or money in the change?

    ★ The Executive Board, the project team, representatives from the plant, and all plant employees.

The project will have a Steering Committee who will be responsible for making decisions, throughout this process, based on recommendations by the project team. The entire Steering Committee has not yet been named; however, it will comprise employees in various job roles.

The Plant Manager, the Executive Board's VP of Finance, the Project Manager, and the Change Management Lead are to be included in the Steering Committee. The Executive Board has indicated that others will be included as well. That being said, Shane Technologies' CEO will approve all purchasing decisions. All other decisions will be voted on by the Steering Committee.

Sample Change Assessment (continued)

6. Who will make decisions along the way?
   - ★ Steering committee
   - ★ Project team
7. Who will approve the decisions?
   - ★ Purchasing decisions: Shane's CEO
   - ★ All other decisions: Steering committee

This concludes Chapter 6: Explain Change and Identify the Motivation Behind an Upcoming Change.

CHAPTER

CREATE A VISION FOR THE FUTURE CULTURE

7

In this chapter, you will learn about creating a vision and its impact on the organization's culture.

Part 2 Outline

A vision is a written statement describing the desired future state of your team or organization after a change initiative.

The process of creating a vision helps to flesh out the project definition before the team begins collecting detailed requirements for the project.

Chapter 7: Create a Vision

❑ Create a vision for the future culture

❑ What is a vision?

⇨ A written statement describing the desired future state of your team or organization after a change initiative

---

There are several benefits of developing a vision for change. Please take a moment to review the items listed in the left column.

On the flip side, one should be on the lookout for vision killers. Vision killers are modes of thought and behaviors that cause an organization to maintain the status quo. Please take a moment to review the items listed in the right column.

Develop a Vision

Benefits	Vision Killers
❑ Identifies direction & purpose	❑ Tradition
❑ Alerts stakeholders to needed change	❑ Fear of ridicule
❑ Promotes interest and commitment	❑ Stereotypes of people, conditions, roles, and governing councils
❑ Encourages openness to unique and creative solutions	❑ Complacency of some stakeholders
❑ Builds loyalty through involvement (ownership)	❑ Fatigued leaders
❑ Results in efficiency and productivity	❑ Short-term thinking
	❑ "Naysayers"

Here's the eHam.net Vision Statement: To build the largest and most complete Amateur Radio community site on the Internet.

This "portal" will be the first place to go for information, to exchange ideas, and be part of what's happening with Ham radio on the Internet. The project will use a management team of volunteers who each manage a topic of interest with passion. The site will stand above all other Ham radio sites by employing the latest technology and professional design and programming standards, developed by a team of community programmers who contribute their skills to the effort. Everyone involved in the development of the site will be proud to say they were a part of the process.

Sample Vision Statement #1

❑ eHam.net

⇨ To build the largest and most complete amateur radio community site on the Internet

⇨ To attract users who use eHam as their source for informaton

⇨ This project involves a management team of volunteers, who each take a topic of interest and manage it with passion

⇨ The site will stand above all other ham radio sites by employing the latest technology and professional design/ programming standards

---

Here's Greif's Five-Year Vision Statement:

We will achieve superior return on assets while consistently increasing revenue.

We will strive to be one of the most desirable companies to work for in our industries, focusing on establishing a work atmosphere in which our employees can excel.

We will continually add value to our customer relationships, with an emphasis on providing packaging solutions on a global scale.

We will capitalize on our leadership position and pursue profitable growth in the Industrial Packaging & Services business.

We will pursue internal growth in addition to strategic acquisitions to profitably expand the Paper, Packaging & Services business.

We will grow our Timber business both in terms of land holdings and revenue generation.

Sample Vision Statement #2

❑ Grief's Five-Year Vision Statement

⇨ We will achieve superior return on assets while consistently increasing revenue

⇨ We will strive to be one of the most desirable companies for which to work

⇨ We will continually add value to our customer relationships

⇨ We will capitalize on our leadership position and pursue profitable growth in the Industrial Packaging & Services business

⇨ We will pursue internal growth

⇨ We will grow our Timber business both in terms of land holdings and revenue generation

---

Here's Shane Manufacturing's vision statement: In 2011–2012, Shane Manufacturing will implement a new system in parts plants that will automate portions of the parts construction process. The new system will replace current processes with new processes that will lower risk of injury for employees and pay for itself within the next three years due to increased production. This project will be managed in a way that will ensure that all employees have the information and training they need to be successful using the new process.

Sample Vision Statement #3

❑ Shane Manufacturing's Vision:

⇨ In 2011-2012 the organization will implement a new system to automate portions of the parts construction process

⇨ The new system will lower the employees' risk of injury

⇨ The new process is expected to pay for itself within three years due to increased production

⇨ Employees will receive necessary information and training to be successful

Write a vision statement for your own company.

Here are five keys you could use to create an attainable vision for your organization. Let's take a look at each one.

Key #1 – Involve the right people. Invite a representative from each functional area that the change is likely to touch. Aim to fill the team with optimistic, energetic, and practical individuals who want to be involved.

Vision Creation

❑ **Key #1: Involve the right people**

❑ Key #2: Be open, flexible about change

❑ Key #3: Consider the present

❑ Key #4: Capture the vision (early stages of the process)

❑ Key # 5: Keep the vision alive

Key #2 – Be open. Imagine a future where changes are implemented smoothly. List the overall goals of the change effort. Be receptive to dramatic change. Describe the future simply and clearly. Be visual. Be inspiring. Be specific.

Vision Creation

❑ Key #1: Involve the right people

❑ **Key #2: Be open, flexible about change**

❑ Key #3: Consider the present

❑ Key #4: Capture the vision (early stages of the process)

❑ Key # 5: Keep the vision alive

Key #3 – Consider the present. List strengths that should remain, and issues that indicate improvement is needed. List the people involved in and affected by the change.

Vision Creation

❑ Key #1: Involve the right people

❑ Key #2: Be open, flexible about change

❑ **Key #3: Consider the present**

❑ Key #4: Capture the vision (early stages of the process)

❑ Key # 5: Keep the vision alive

Key #4 – Capture the vision. A vision can be one or two paragraphs or one or two pages long. Revise the vision if necessary in the early stages of the project, however, do not revise after your solution is designed.

Vision Creation

- ❑ Key #1: Involve the right people
- ❑ Key #2: Be open, flexible about change
- ❑ Key #3: Consider the present
- ❑ **Key #4: Capture the vision (early stages of the process)**
- ❑ Key # 5: Keep the vision alive

Key #5 – Keep the vision alive. Keep the vision document visible throughout the project. The team should continuously measure their actions and results against the overall project vision.

Vision Creation

- ❑ Key #1: Involve the right people
- ❑ Key #2: Be open, flexible about change
- ❑ Key #3: Consider the present
- ❑ Key #4: Capture the vision (early stages of the process)
- ❑ **Key # 5: Keep the vision alive**

This concludes Chapter 7: Create a Vision for the Future Culture.

DETERMINE HOW CHANGE WILL IMPACT PROCESSES AND PEOPLE

8

In this chapter you'll apply what you learn by identifying the necessary ingredients for a successful change and completing a training action plan.

These are the chapters that we'll cover during the section. Remember to keep your pen and paper handy to take notes or to try the exercises.

    In this section you will determine how change will impact processes and people; identify the key processes that will be impacted by the change initiative; analyze groups of people and how they are affected; and assess the impact of the change initiative on the overall organization.

Process Improvement

❑ Determine how change will impact processes and people

❑ Identify the key processes that will be impacted by the change initiative

❑ Analyze stakeholders and how they are affected

❑ Assess the impact of the change initiative on the overall organization

---

PROCESS IMPROVEMENT

How do you do define process? How do you define improvement?

Defining Process Improvement

❑ How do you do define process?

❑ How do you define improvement?

A process is a series of actions, changes, or functions bringing about a result. It could also be a series of operations performed in the making or treatment of a product; or a series of progressive and interdependent steps by which an end is attained.

    Improvement is the promotion in desirable qualities, progress toward what is better, or the act of making profitable use or application of anything.

Defining Process Improvement

❑ How do you do define process?

⇨ A series of actions, changes, or functions bringing about a result

⇨ A series of operations performed in the making or treatment of a product

⇨ Progress; passage

❑ How do you define improvement?

⇨ Promotion in desirable qualities; progress toward what is better

⇨ The act of making profitable use or application of anything

---

Process improvement can be defined as designs or activities, which improve quality or reduce costs, often through the elimination of waste or non-value-added tasks.

Defining Process Improvement

❑ How do you do define process improvement?

⇨ Designs or activities, which improve quality or reduce costs, often through the elimination of waste or non-value-added tasks

---

Process Improvement is related to Change Management because when processes are changed people are affected.

When processes are changing you should answer the following questions to keep track of how various aspects of the organization are affected:

What are the current processes?
How will current processes differ from the future state?
What impact will changes to processes have on jobs?
What roles will change and how will they change?
What skill sets are required?
What are the new job descriptions?

Identifying Processes That are Changed

❑ When processes are changing you should be on the lookout for the following:

- ⇨ What are the current processes?
- ⇨ How will current processes differ from that of the future?
- ⇨ What impact will changes to processes have on jobs?
- ⇨ What roles will change and how will they change?
- ⇨ What skill sets are required?
- ⇨ What are the new job descriptions?

CASE STUDY

Grede Foundries, Inc., headquartered in Milwaukee, WI, is a fast-growing foundry operator, adding 6 new plants since 1984. Founded in 1920, the company now operates a total of 11 foundries in 8 states, which produce gray iron, ductile iron and steel castings.

Nearly half of its cast products are used in the automotive industry. Other markets consist of machine shops and a range of additional industries, including construction equipment, farm equipment, pumps, mechanical power equipment, hydraulic valves, engines and air conditioning.

The company is now also extending its business reach overseas, through direct exports and technical alliances with foreign foundries in Europe and the Far East.

To maintain its growth in a highly competitive industry, Grede Foundries invests heavily in both new foundry equipment and employee training. The company is also keenly aware of the need for continuous process improvement. "Since almost 50% of our business is with the auto industry, we have to respond to the same demands they face relative to price and quality," says Chris Paras, industrial engineer at Grede Foundries' Milwaukee office.

> ## Process Improvement Case Study
>
> ❑ Case Study: Simulation of Foundry Operations Yields
> Productivity Gain at Grede Foundries, Inc.
>
> ⇨ Nearly half of Grede Foundries' cast products are used in the
> automotive industry
>
> ⇨ The company is now also extending its business reach
> overseas, both through direct exports and technical alliances
> with foreign foundries in Europe and the Far East
>
> ⇨ Large investments in new equipment and employee training to
> maintain growth in its competitive industry
>
> ★ This allows Grede to be flexible, and meet customers'
> needs and demands relative to price, quality, and changes
> to product specifications

An important step in this direction was the acquisition of
QUEST discrete event simulation software in 1995. According
to Paras, the technology was purchased to handle "what if"
questions—to define potential problems or bottlenecks in a
foundry system before hiring extra labor, adding equipment, or
altering processes.

"We've been using the simulations to evaluate our processes," Paras says. "That covers operations within the work-
cells, the transport conveyors, AGVs and some labor, too. We
are trying different scenarios to learn how we can operate best
to meet current demands, and also how we can meet long-term
goals within budget."

To answer these questions, Paras does simulations at the
company's eleven foundries, heading teams that usually include
the plant's facilities engineers, industrial engineers, and area
managers. "It was tough getting buy-in at the beginning," he
says, "but I've learned to work around that in a couple ways.

First, I get the other people involved with the data collection,
and then again during the validation/verification phases. That
allows me to demonstrate that the model we're working with
accurately reflects what's happening on the shop floor. After I've
proven to them that the base model is correct, I'll go ahead and try
different alternatives. At that point, buy-in is no longer a problem."

Process Improvement Case Study (continued)

❏ Case Study: Grede Foundries, Inc.

⇨ Acquisition of QUEST discrete event simulation software to help the company respond to identify and respond to potential problems or bottlenecks

* This helps the company make decisions involving hiring, purchasing equipment, or altering processes

* Use the scenarios to determine how to meet demands, while staying on budget

⇨ Include employees in the process via demonstrations

What is the main process improvement, which resulted from the implementation of the new software?

Grede implemented QUEST software in order to simulate multiple processes, which look for opportunities for improvement. The main process improvement is that Grede implemented a continuous process to monitor efficiency.

How did Grede manage the change? The company managed this large change in the way it did business, by involving employees, who were affected by the changes, early and often.

Process Improvement Case Study: Reflection

❏ What is the main process improvement, which resulted from the implementation of the new software?

❏ How did Grede manage the change?

STAKEHOLDER ANALYSIS

A stakeholder is a person or group who can affect or is affected by an action. The Stakeholder Analysis is a tool that is used to map out the people and impact of change. It addresses or ameliorates stakeholder concerns, and is the means by which communication and training are delivered throughout the change process.

The Stakeholder Analysis includes the stakeholders, their respective concerns, key messages and actions, and vehicles used for communication and training.

The Stakeholder Analysis contributes to project design by identifying the goals and roles of different groups, and by helping to formulate appropriate forms of engagement with these groups. One of the keys to an effective Stakeholder Analysis is to ensure that you brainstorm with every group touched by the change, even if it is affected in a minor way.

Stakeholder Analysis

❑ Who is a stakeholder?

⇨ A **stakeholder** is a person or group who can affect or is affected by an action

❑ Stakeholder Analysis is a tool that is used to map out:

⇨ **Stakeholder**= The categories of people impacted by the change

⇨ **Change Impact**= Stakeholders' main concerns related to the change

⇨ **Key Messages/ Actions**= What should be done to address stakeholders' concerns

⇨ **Vehicle**= The communication/ training medium preferred by the group

---

Here is a sample Stakeholder Analysis used for a company that is implementing a Skills Inventory. Review how the change being implemented impacts the various stakeholder groups in the organization. Let's look at one of the analyses.

Managers and supervisors are impacted by this change like any employee. They might feel uncertain about their

respective positions in the company. They might ask themselves, "do I really have the required skills for this job?"
They are responsible for how the change impacts their direct
reports as well.

The key actions which the organization must take are to
ensure that managers understand how and why the inventory was developed and how it will benefit all employees in
the organization. Regular training seminars are the best vehicles for engaging managers and increasing the likelihood of
buy-in.

Sample Stakeholder Analysis

Stakeholder	Change Impact	Key Messages/Actions	Vehicles
All Officers	• They are going to have to explain Skills Inventory process to employees – and they've never had to explain anything like this before • They may not really understand why we're doing this • They need to spend some time on/support the project – by answering employee questions, encouraging employees to take the survey/managers to verify, etc.	• Clearly explain the project and objectives to them • Describe the benefits of the project • Ask them to encourage managers to verify their employees' surveys • Provide them with speaking points so they know what to say about the project (pending the change management activities that are included in the plan)	• Meetings • Emails • Information Packets
Managers/ Supervisors	• All employee impact (below) points apply here • Responsible for discussing survey with employees • Responsible for reviewing survey • Responsible for having conversations with employees if there are issues with the survey • May have interest in reporting	• Overview of the program and this year's skills survey • Program Objectives • High-level Project Activities/Timeline • Amount of time this will take for them/employees • How skills lists were created • How this is different from the annual review • CM Team's expectations for conclusions employees might draw about survey without discussions from supervisors	• Emails • Seminar Style "Manager Only" Training • Information Packets

Here, you can review how all employees may perceive the
change's impact on their jobs and how the organization can
ameliorate concerns.

Sample Stakeholder Analysis (continued)

Stakeholder	Change Impact	Key Messages/Actions	Vehicles
All Employees	• Fear of losing job/layoff • Not understanding what is expected • Length of time it takes to complete survey • Instructions on how to take the survey • Understanding the reported results • Feelings of insecurity if they only have a few skills	• Articulate objectives of the survey (i.e., staff cross-functional project teams) • Objective assessment of strengths and areas for professional development • Clarify and standardize expectations • Provide criteria for success advancement, career development	• Emails • Meetings with direct supervisor • Articles on Corporate Intranet

ASSESSING CHANGE IMPACT

Once you have discovered and documented the processes that are changing, and your stakeholder groups take time to "score" each process and stakeholder group, you can prioritize activities in your plan for both process change and behavior change.

Assessing Change Impact

❑ Prioritize processes and actions

⇨ **Process Change:** Changes to procedures, tools, order of events, etc., within a process

⇨ **Behavior Change:** Changes to the manner in which people are required to act, or react in a situation

By using the Process Change Matrix, you can identify which process changes are most important. This is a key step in planning. Writing the plan allows you to determine which activities should receive the most resources (time, money, people, etc.) and which might be less important in the big picture.

This is an example of a process change matrix. It allows one to categorize the activities, which require varying degrees of behavior changes, cross-matched with the expected degree of resistance to that change.

The matrix demonstrates how one could choose to allocate resources to different stakeholders. For example, one department will need to undergo the greatest shift in change and responsibilities, necessitating the greatest allocation of resources to support the change.

The matrix focuses on the initiative and required behaviors needed to change, and does not refer to specific stakeholders.

In this sample process matrix, you can see that Shane Manufacturing expects that the new manufacturing process will demand the most amount of resources because a new technology is being introduced to record data.

The number of steps needed to complete the process are going to be changed and reduced. This change would likely require frequent communication, training, and reinforcement for stakeholders.

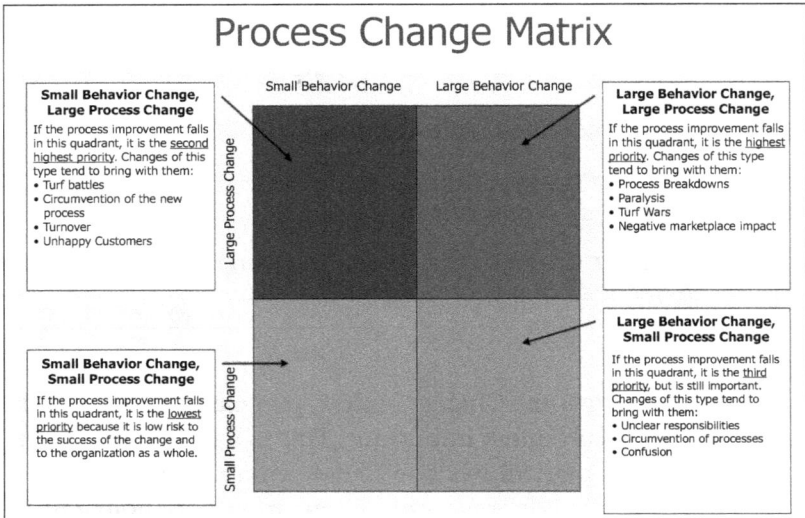

Process Change Matrix

Small Behavior Change   Large Behavior Change

Small Behavior Change, Large Process Change

If the process improvement falls in this quadrant, it is the second highest priority. Changes of this type tend to bring with them:
• Turf battles
• Circumvention of the new process
• Turnover
• Unhappy Customers

Large Behavior Change, Large Process Change

If the process improvement falls in this quadrant, it is the highest priority. Changes of this type tend to bring with them:
• Process Breakdowns
• Paralysis
• Turf Wars
• Negative marketplace impact

Large Process Change

Small Behavior Change, Small Process Change

If the process improvement falls in this quadrant, it is the lowest priority because it is low risk to the success of the change and to the organization as a whole.

Small Process Change

Large Behavior Change, Small Process Change

If the process improvement falls in this quadrant, it is the third priority, but is still important. Changes of this type tend to bring with them:
• Unclear responsibilities
• Circumvention of processes
• Confusion

Process Change Matrix for Shane Manufacturing

Process Change	Description	Large or Small Behavior Change? (L,S)	Large or Small Process Change? (L,S)	Priority L+L = 1 S+L = 2 L+S = 3 S+S = 4
New Process for manufacturing car part #1	The original process for manufacturing car part #1 had 12 steps. The new process will have 8 steps, and 4 of the steps require that employees use the new technology to record data.	L	L	1
New Process for packaging parts	We are changing the supplier for our packaging. The new packaging eliminates the need to use a glue gun. Now our boxers will use the box sealing tape that is included as part of each box speeding the process.	S	S	4
The Order Routing/Purchasing department merger	Because of new efficiencies, the Purchasing department and the Order Rounding departments will be merged. No processes will change for Order Routing.	S	L	2

Let's review the stakeholder matrix. Notice that this Matrix refers to stakeholder groups, rather than specific processes. It focuses on the level of resistance to change, rather than how large the change is. For example, a change can be relatively large, such as using a new technology. However, a stakeholder group may be eager to use it, demonstrating a low resistance to change.

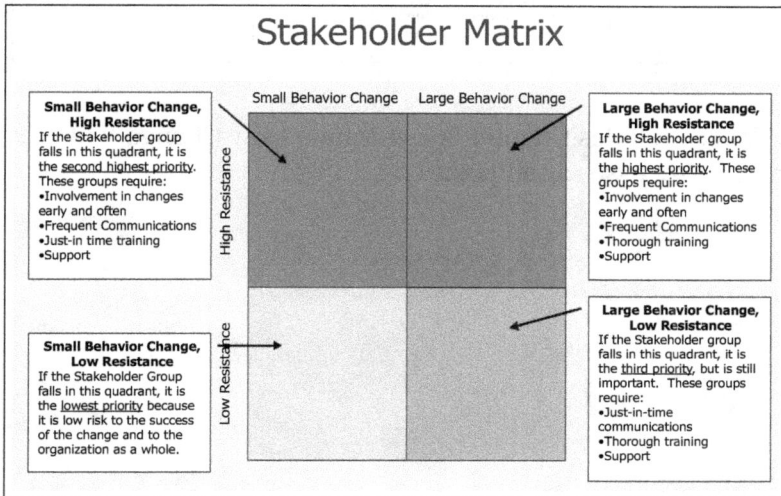

Stakeholder Matrix

Small Behavior Change, High Resistance
If the Stakeholder group falls in this quadrant, it is the second highest priority. These groups require:
•Involvement in changes early and often
•Frequent Communications
•Just-in time training
•Support

Large Behavior Change, High Resistance
If the Stakeholder group falls in this quadrant, it is the highest priority. These groups require:
•Involvement in changes early and often
•Frequent Communications
•Thorough training
•Support

Small Behavior Change, Low Resistance
If the Stakeholder Group falls in this quadrant, it is the lowest priority because it is low risk to the success of the change and to the organization as a whole.

Large Behavior Change, Low Resistance
If the Stakeholder group falls in this quadrant, it is the third priority, but is still important. These groups require:
•Just-in-time communications
•Thorough training
•Support

Here's the Shane Manufacturing Stakeholder Matrix. Notice that the union employees are expected to require the most resources in order to ensure that these stakeholders successfully change. The analysis has used past behavior to determine that this group would likely adapt to change slowly and be resistant to initiatives that Shane implements.

Stakeholder Matrix for Shane Manufacturing

Stakeholder Group	Description	Large or Small Behavior Change? (L,S)	Large or Small Process Change? (L,S)	Priority L+L = 1 S+L = 2 L+S = 3 S+S = 4
Union Employees	Union employees will have highly changed jobs due to the new processes and new technology. Historically, this group at Shane Manufacturing is slow adapting to the idea of new ways of doing business and exhibits high stress levels related to change.	L	L	1
Floor Management	Floor Management is responsible not only for learning their new responsibilities, but also for assisting their employees through the transition. They are open to change.	L	S	3
Executive Board	The Executive Board is fully behind this initiative. They will not see significant change to their jobs although they will be very interested in what is going on with this project (on a high level).	S	L	2

This concludes Chapter 8: Determine How Change Will Impact Processes and People.

IDENTIFY EMPLOYEE DEVELOPMENT NEEDS

9

In this chapter you'll apply what you learn by identifying the necessary ingredients for a successful change and completing a training action plan. These are the chapter that we'll cover during the section. Remember to keep your pen and paper handy to take notes or to try the exercises.

Part II Outline

In this chapter you will learn how to assess the readiness of an organization to buy-in to changes leadership wants to implement. You will also learn how to identify creative communication and training methods.

Chapter 9: Identify Employee Development Needs

❑ Assess the readiness of the organization

❑ Identify creative communication and training methods

ASSESSING ORGANIZATIONAL READINESS

Before a team or company makes a change, it is important to understand the strengths and weaknesses of the organization as they pertain to the upcoming change. An Organizational Change Readiness Assessment is a tool that you can use to measure if a company is ready, willing, and able to make the change happen and succeed.

There are many tools that can be used to gather readiness data. These tools can be used either alone or in combination. Ideally, several approaches are used because each has its benefits and drawbacks.

Assessing Organizational Readiness

❑ Assessment tools:
   ⇨ Surveys & quizzes
   ⇨ Facilitated meetings
   ⇨ Other meetings
      ★ Town Hall meetings
      ★ Staff meetings
      ★ Presentations
      ★ Sales Meetings
      ★ Informal Conversations
      ★ Training

SAMPLE ORGANIZATIONAL READINESS CHART

This chart shows the various tools that organizations use to determine if the systems that are in place and current employees are ready for new initiatives. Each tool has benefits and drawbacks. More than one tool may be required in order for the data to be reliable and valid.

Assessing Organizational Readiness

Tool	Description	Method Benefits	Method Drawbacks
Surveys Quizzes	Surveys that gather information can be purchased or created in house. Quizzes measure understanding	• Anonymity • Measurable data • Large amount of data with relatively little effort	• No ability to probe respondents for reasons behind answers • Certain categories of stakeholders may not complete survey • Requires resources and planning
Facilitated Meetings	A neutral person (consultant, Human Resources) facilities meetings with groups of employees that represent a cross-section of all of the stakeholders impacted.	• Measurable data • Ability to probe respondents for reasons behind answers	• Must obtain time from stakeholders for participation • Participants might be hesitant to share thoughts • Requires resources and planning
Other Meetings: • Town Hall Meetings • Staff Meetings • Board Meetings • Presentations • Sales Meetings • Informal Conversations • As a part of training	Any interaction that involves stakeholders is a time that readiness data can be collected.	• Can be performed on an ad hoc basis • Ability to probe respondents for reasons behind answers	• Data is difficult to measure • Participants might be hesitant to share thoughts

CHANGE READINESS SCORECARD

Even more important than the tool used to collect data, is the type of information you gather. This is a sample of statements used in a Change Readiness Scorecard.

Users are asked to rate their organization on a scale of 1 to 7 next to each of the 17 items. The tool calculates the total score, and determines if the organization is ripe for a change initiative. That is, are employees and systems capable of a successful change initiative.

Change Readiness Scorecard

1. We have a very clear understanding of the likelihood of success with this change.

2. Our business case for change has been broadly communicated and accepted.

3. Our organizational leaders are clear about the leadership behaviors they need to be successful and are actively working to perform more effectively.

4. We have clear metrics for measuring the results of our change initiative.

5. We have a change management plan that integrates the process, technology and people aspects of the change and allows for course correction.

6. We have analyzed all of the possible people impacts and have a plan to address them.

ORGANIZATIONAL READINESS REPORT

Although at first glance it looks like employees are confident because of the good numbers in Agree and Strongly Agree, those numbers only represent 42% of employees. Additional training may be needed. The Support team needs to be aware of the low confidence rate, as well.

Sample Organizational Readiness Report

I feel confident that I will be successful using the new process.		
Response	**% of Respondents**	**Red Flag Issue?**
Strongly Disagree	20%	
Disagree	26%	🏳
Neutral	12%	
Agree	34%	
Strongly Agree	8%	

According to this report, Shane Manufacturing leadership is doing an excellent job at sponsoring the upcoming change. Leaders should continue to sustain this level of sponsorship through the launch and implementation in order to ensure success.

Sample Organizational Readiness Report

❑ Is the organization's leadership visibly supportive of the new initiative?

Our organizational leaders visibly demonstrate commitment to the change in both their words and behaviors.		
Response	**% of Respondents**	**Red Flag Issue?**
Strongly Disagree	8%	
Disagree	12%	
Neutral	8%	
Agree	40%	
Strongly Agree	32%	

Take a moment to think about all of the communication vehicles and training methods that your organization uses and ones that you think that should be used.

Communication & Training

❑ Think!

⇨ List communication vehicles

⇨ List training methods

Here are some example of communication and training vehicles. These are not exhaustive lists, and you should feel free to add more.

Communication & Training Vehicles

❑ Communication Vehicles   ❑ Training Vehicles

⇨ Presentation	⇨ Interactive website
⇨ Organization-wide meetings	⇨ Self-directed learning
⇨ Organization-wide voice mail	⇨ Instructor-led training
⇨ Staff meetings	⇨ Online training
⇨ Focus groups	⇨ Web-conferencing
⇨ One-on-one meetings	⇨ E-learning
⇨ Email	⇨ On-the-job training
⇨ Informal conversations	⇨ Training Manual
⇨ Newsletters	⇨ Job-Aid (one-two page
⇨ Status reports	training document)
⇨ Corporate Intranet	
⇨ Memos/Flyers	
⇨ Screen-savers	

In this case the implementation of a new CRM tool was not just a technology roll-out, but a culture change initiative, especially for the sales force. The company needed to shape the project's implementation in terms of people, processes, and technology.

Though the organization was reluctant to change, the project might not have progressed without strong business support at the executive level. The division head as well as top executives insisted that all CRM users be trained as soon as possible to provide the best possible customer service.

The president of sales accepted sales forecasting information through the new CRM tool to compel the entire sales force to rely on the new system for all data entry points of the sales cycle.

Case Study

❑ Communication & Training: Addressing People Issues in a Customer Relationship Management (CRM) Implementation

    ⇨ **It starts with strong business sponsorship**

    ⇨ Train everyone

    ⇨ Provide follow-up support

    ⇨ Communicate often and proactively in many media

The CRM training took several forms. First, everyone in the entire organization received an overview of the CRM tool, learning how it would help in their job.

CRM users also attended either a hands-on training session or a webcast. Users also received a follow-up training tool, prior to the system's launch.

The organization also offered coaching, additional training, support, and hand holding. The personal assistance available from the project team was critical to getting users over the learning curve and correcting software errors.

Case Study

❑ Communication & Training: Addressing People Issues in a Customer Relationship Management (CRM) Implementation

⇨ It starts with strong business sponsorship

⇨ **Train everyone**

⇨ Provide follow-up support

⇨ Communicate often and proactively in many media

For projects involving a culture change, the value of proactive communication cannot be overstated. The communication plan identified the major stakeholders, their needs and ways to satisfy those needs.

Another essential part of the plan was to encourage communication from employees. For example, users were able to suggest improvements or submit enhancement requests. The organization also used several communication vehicles, including status reports, email updates, monthly announcements of business accomplishments, and reminders of training sessions. The communication plan successfully maintained employee buy-in throughout the process.

This concludes Chapter 9: Identify Employee Development Needs.

WRITE A PLAN FOR MOBILIZING AN ORGANIZATION

10

In this chapter you will learn how to write a plan for mobilizing an organization, applying what you learn on the job and identifying the necessary ingredients for a successful change.

By the end of this section, you will be able to explain "Change;" identify the motivation behind change; create a vision for the future; determine how a change will impact processes and people; identify employee development needs; and write a plan for mobilizing an organization.

To get the most out of this section, you should identify a change that will affect you, your team/department or your organization. As the chapters progress, consider how to apply the principles to your situation. You should have a pen and paper handy to take notes or try exercises on your own.

Learning Objectives

By the end of this section, you will be able to:

- ❑ Explain "Change"
- ❑ Identify the motivation behind an upcoming change
- ❑ Create a vision for the future culture
- ❑ Determine how a change will impact processes and people
- ❑ Identify employee development needs
- ❑ Write a plan for mobilizing an organization

In this section you'll apply what you learn by identifying the necessary ingredients for a successful change and completing a training action plan. These are the chapters that we'll cover during the section. Remember to keep your pen and paper handy to take notes or to try the exercises.

Part 2 Outline

Chapter 6:   Explain "Change" & Identify The Motivation Behind An Upcoming Change

Chapter 7:   Create A Vision For The Future Culture

Chapter 8:   Determine How A Change Will Impact Processes And People

Chapter 9:   Identify Employee Development Needs

Chapter 10: Write A Plan For Mobilizing An Organization

Now you will learn how to write a plan for mobilizing an organization, apply what you learn on the job and identifying the necessary ingredients for a successful change.

HERE IS A SAMPLE PROJECT PLAN
REVIEW ITS COMPONENTS

Sample Project Plan

Segment/Task/Activity	Estimated Effort (In Days)	Target Date	Responsibility	Status	Comments
Project Setup, Administration, and Mgmt.				**Segment Owner--SO**	
Hold team kick-off meeting	0.3	2/25/2003	Project Team	C	
Project team status meetings	17.0	Mondays and Thursdays of each week	DP, AR, SO, JM, TB	On-going	
Analysis				**Segment Owner--SO**	
Draft Stakeholder Analysis	1.5	2/12/2003	SO	C	
Determine strategy for involving various levels of leadership	0.3	2/24/2003	Project Team	C	
Meet with Executive Officers to gather information about vision	0.5	2/25-3/3	ST, SO	IP	
Revise project design document	0.5	3/5/2003	SO		Will update document based on feedback from Executive Officers.
Communications				**Segment Owner--JM**	
Send all employee initial communication to all employees (electronic and paper)	0.1	5/29/2003	CEO	IP	
Draft speaking points for Jim to present at officer mtg	0.3	5/1/2003	JM		
Project updates to Senior Management	2	Bi-weekly from 5/19/03 - 8/1/03	JM		Short. Delivered through email. Concentrates on information that should be disseminated down through Senior Manager's org.
Draft Newsletter Update	0.5	Monthly from May - Aug	JM		

INGREDIENTS FOR SUCCESSFUL CHANGE

These are the key ingredients one must have to implement a successful change in an organization. Please consider each one. Then, rate your organization, on a scale of 1-5, with 5 being the highest.

Let's start with **#1, which is Leadership**: "Our leaders sponsor change with their vocal and visible support. Our Leaders can challenge, motivate, and empower teams through change."

Key Ingredients for Successful Change

❑ What are the key ingredients to successfully change an organization?

1. Leadership

2. Vision

3. Integrity

4. Involvement

5. Accountability

6. Communication

7. Acknowledgement

8. Progress/Completion

9. Celebration

Next, consider **#2 Vision**: "We have a written vision statement describing the desired future state of the team or organization after the change initiative."

Key Ingredients for Successful Change

❑ What are the key ingredients to successfully change an organization?

1. Leadership

2. Vision

3. Integrity

4. Involvement

5. Accountability

6. Communication

7. Acknowledgement

8. Progress/Completion

9. Celebration

#3 is Integrity: "Our leadership possesses a high-level of personal integrity – they are honest and moral."

Key Ingredients for Successful Change

❑ What are the key ingredients to successfully change an organization?

1. Leadership
2. Vision
3. **Integrity**
4. Involvement
5. Accountability

6. Communication
7. Acknowledgement
8. Progress/Completion
9. Celebration

#4 Involvement: "Our change initiative involves everyone impacted by the change. Each stakeholder group has a representative that has a voice. Individual employees have the ability to provide input and feedback in a way that gives people a sense of ownership."

Key Ingredients for Successful Change

❑ What are the key ingredients to successfully change an organization?

1. Leadership
2. Vision
3. Integrity
4. **Involvement**
5. Accountability

6. Communication
7. Acknowledgement
8. Progress/Completion
9. Celebration

#5 is Accountability: "People at our company feel a part of our successes, feel they are involved, have the ability to fail, and make corrections as necessary."

Key Ingredients for Successful Change

❑ What are the key ingredients to successfully change an organization?

1. Leadership
2. Vision
3. Integrity
4. Involvement
5. Accountability

6. Communication
7. Acknowledgement
8. Progress/Completion
9. Celebration

#6 is Communication: "We communicate changes early and often, using multiple communication vehicles. We have methods in place for gathering feedback from those impacted by the change."

Key Ingredients for Successful Change

❑ What are the key ingredients to successfully change an organization?

1. Leadership
2. Vision
3. Integrity
4. Involvement
5. Accountability

6. Communication
7. Acknowledgement
8. Progress/Completion
9. Celebration

#7 is Acknowledgement: "Our organization acknowledges both successes and setbacks, and communicates them with employees."

Key Ingredients for Successful Change

❑ What are the key ingredients to successfully change an organization?

1. Leadership
2. Vision
3. Integrity
4. Involvement
5. Accountability

6. Communication
7. **Acknowledgement**
8. Progress/Completion
9. Celebration

#8 is Progress & Completion: "We have measures in place that show how progress is being made to the final goal."

Key Ingredients for Successful Change

❑ What are the key ingredients to successfully change an organization?

1. Leadership
2. Vision
3. Integrity
4. Involvement
5. Accountability

6. Communication
7. Acknowledgement
8. **Progress/Completion**
9. Celebration

And finally, consider **#9, which is Celebration**: "We take the time to celebrate our successes."

Key Ingredients for Successful Change

❑ What are the key ingredients to successfully change an organization?

1. Leadership
2. Vision
3. Integrity
4. Involvement
5. Accountability

6. Communication
7. Acknowledgement
8. Progress/Completion
9. Celebration

Implementing change in an organization can be a painstaking process. The organization must consider all of the business process that will be altered, removed, and added as well as how these changes will impact the workforce. Remember, a significant transformation to how the organization functions will have an impact on its culture.

Change affects stakeholders in the organization from officers to entry-level employees. Frequent, honest, and clear communication to stakeholders is vital to the success of this process. They need to know what, why, when, and how something is going to happen.

Employees will always wonder, "What's in it for me? Why should I care?" Remember, individuals adapt to change over four stages: awareness, understanding, acceptance, and commitment.

How well an organization manages this process, using communication, training, and support tools, affects how well employees will move through it. Will talented employees thrive or will the organization experience high attrition rates?

Conclusion

❑ Implementing change in an organization can be a painstaking process, which must consider several factors

❑ Change affects stakeholders in the organization from officers to entry-level employees

❑ Frequent, honest, and clear communication to stakeholders is vital to the success of this process

  ⇨ Overcome employee resistance to change

  ⇨ Understanding that individuals move through a series of stages after a change is announced

Take a look at these websites for additional reference material

References

❑ Organizational Change Readiness Score Card

  ⇨ http://www.darkhorsestrategies.com/consult/organizationalchange.htm

❑ The Definition and History of Change Management: Change Management Tutorial Series

  ⇨ http://www.change-management.com/

❑ Wojtecki, J.G. and Peters, R.G. (2000). Communicating Organizational Change: Information Technology Meets the Carbon-Based Employee Unit. *The 2000 Annual Consulting, Vol. 2*, 175–190. San Francisco, CA: Jossey-Bass/Pfeiffer.

  ⇨ http://www.CenterForRiskCommunication.com/pubs/crc-p3.pdf

❑ Hiatt, J. "Five Steps for Creating Effective Visions."

  ⇨ http://www.prosci.com/visions.htm

This concludes Chapter 10: Write a Plan for Mobilizing an Organization.

www.ingramcontent.com/pod-product-compliance
Lightning Source LLC
Chambersburg PA
CBHW061830220326
41599CB00027B/5244